Mathematics

THE PRODUCTIVE STRUGGLE

Alvin Allen
and
Alva White

authorHOUSE®

AuthorHouse™
1663 Liberty Drive
Bloomington, IN 47403
www.authorhouse.com
Phone: 833-262-8899

Published by AuthorHouse 03/19/2021

ISBN: 978-1-6655-1943-4 (sc)
ISBN: 978-1-6655-1945-8 (e)

Library of Congress Control Number: 2021904987

Print information available on the last page.

This book is printed on acid-free paper.

About the Authors

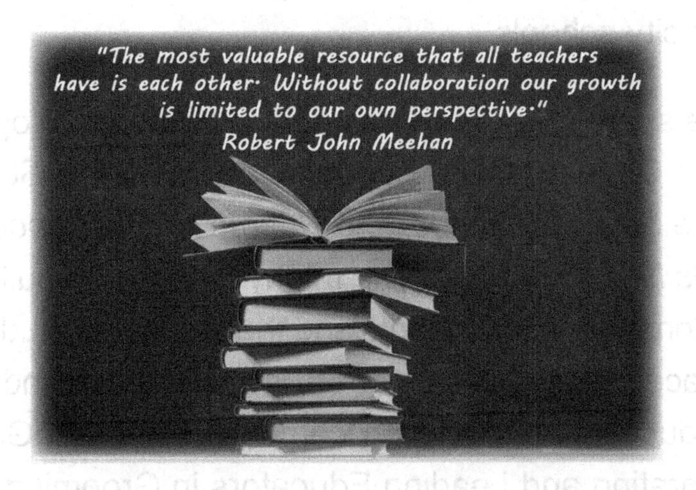

"The most valuable resource that all teachers
have is each other. Without collaboration our growth
is limited to our own perspective."
Robert John Meehan

MEET YOUR AUTHORS, ALVIN ALLEN and ALVA WHITE. Both are educators that have a strong love for mathematics. Not for just doing the math but living math. After all, living is learning and effective teachers are lifelong learners.

ALVIN ALLEN has 18 years of experience in the field of education. During his earlier educational travels, he was a middle school and high school teacher, a STEAM coordinator, and district math consultant. Currently, he serves as a Career and Technology Education

Consultant for a school district in Columbia, SC. Other notable achievements of Alvin are his degree in Educational Technology from Webster University, serving as teacher of the year twice during his career, and recipient of the Instructional Technology Spotlight Award. Also, he had a 100% passing rate on the South Carolina Algebra 1 EOC for 7 years at three different inner city schools.

In his spare time, Alvin has self-published a trilogy of books *Capers Middle School, Capers Middle School II: The Saga Continues, and Capers Middle School III: Life Is Not a Fairytale.* The trilogy focuses on building relationships with students and the pressures students are faced within the school. He is also the cofounder of an educational consulting company called iLEGACY (Integrating and Leading Educators in Grooming and Cultivating Youth). Alvin feels that all students deserve the opportunity to learn. He goes by the motto, "Hard Work is the Key to Success!!!!".

ALVA WHITE has 24 years of experience in the field of education. She has served as a middle school teacher, curriculum resource teacher, and currently a math consultant for a school district in Columbia, SC. With a BS in Mathematics from Benedict College, a Masters of Education in Divergent Learning from Columbia College, and her experience with overseeing a youth program for girls, the Betty Shabazz Delta Academy Program, for her chapter, Columbia (SC) Alumnae Chapter of Delta Sigma Theta Sorority, Inc., she always finds herself living life's lessons while impacting learning for others.

Alva has also served the mathematics education profession as an officer and board member for SCCTM (South Carolina Council of Teachers of Mathematics) and SCLME (South Carolina Leaders of Mathematics Education). Alva was the 2019 recipient of the SCCTM Outstanding Contribution to Mathematics Education Award. Her love is to share her knowledge of math with others, students and teachers alike. She truly believes that "Knowledge is POWER."

Introduction

MATHEMATICS
is not about
numbers, equations,
computations, or
algorithms:
it is about
UNDERSTANDING
William Paul Thurston

WHEN WE AS TEACHERS WALK into a math classroom, there are many obstacles to accomplish. Not only do we have to know the depth of each standard but we have to articulate and demonstrate them with clarity to our students. Much of the understanding of the craft of teaching comes through on-the-job training. Your classroom is your domain. You are the king or queen of your domain. You have to know the ins and outs for your domain. You establish the success criteria in your domain. You have to know what works and what needs tweaking, what is real world and what's real in a student's world, and what's the difference between learning for the moment and lifelong learning. Over time and with various experiences, your craft is perfected. Teachers, effective teachers, are always learning.

Mathematics: The Productive Struggle is our effort to share some of our learned experiences that will support you on your journey to equipping your toolbelt with tools that will aid you when working with middle and high school learners in your domain. Any teacher that continues to learn and grow through new experiences will always find new value and be able to address the needs of students that cross the threshold of their door. We are sure you can agree that the students of today are very different than when you were a student in school. Currently, students bring new challenges to the classroom that has to be identified and addressed for true learning to occur. Hence, there is a need to address what we as teachers do through what we have tagged productive struggle.

The productive struggle can take on many looks at different times throughout your career. This is why learning can never come to an end while working in the field of education. We hope that this resource will enter your journey at a time that you are looking to add to or sharpen a tool in your toolbelt. As you read and engage keep in mind that there is not a one size fits all for teaching. Nor is there a strategy that will work in every situation. However, we can ensure that the strategies shared here will ignite a spark that could start that flame in your thoughts and actions that will help you to become the teacher you desire for your students.

THE STARTING POINT
OF ALL ACHIEVEMENT
IS DESIRE
NAPOLEON HILL

Chapter 1: Having a Growth vs. Fixed Mindset In Mathematics 1
"I hate math." "These kids can't do it."

Chapter 2: Relationship Building 5
"You can't teach them, until you reach them."

Chapter 3: Mathematical Discourse 11
"Talk to me, let me know what you are thinking."

Chapter 4: Building Numeracy 19
"The bridge between knowing and applying, that's where you will find numeracy."

Chapter 5: Differentiation in Math Class 25
"Change It, Rearrange It, They Will Bring It"

Chapter 6: From Isolation to Collaboration 31
"We Are In This Together"

Chapter 7: Test Preparation 37
"Get Ready To Show Your Skills"

Chapter 8: Framework For Success 41
"Foundation is Necessary"

Chapter 9: Assessments 51
"ACT - Acknowledge, Create, Time"

Chapter 10: Closure 57
"The Game Doesn't End Until You Close It Out"

Summary 61

Having a Growth vs. Fixed Mindset in Mathematics

"I hate math." "These kids can't do it."

HAVE YOU EVER HEARD THESE words from your students, "I can't do math" or "Math is too hard" or "I know I am going to fail this class"? Honestly, have you ever had these thoughts after looking at data from your students' prior years? "These kids are so far behind they will never be able to do grade level work" or "What am I supposed to do with all these kids that have IEPs?" These are some of the different challenges we face as teachers in the school system. Most teachers ask this question at the beginning of each year. ***How can we change our mindset, as well as the mindset of our students?***

The first step we would suggest in changing your mindset is to take a hard look in the mirror. If you don't feel that all your students can be successful, then you have already created a fixed mindset that failure is an option. You have come to a consensus within you that some of your students will not be successful. These comments are not intended as judgment or criticism. The bottom line is, if there lives an option in our thoughts for a seed to be planted, then there lives the option for that seed to grow. Hence, if there is a slight chance that you think or feel a student cannot be successful, then there is a lens for viewing an action, thought, or missed opportunity as failure. You control the seed that is planted.

Once you have taken a look at yourselves, the next step is developing strategies that will change the mindset of your students.

Step 1: Create a safe and inviting environment for your students to feel comfortable about math. As a teacher, I put long term and short term math goals up on the board. I let the students know that all of us will reach the goals. I also state from day one and throughout the year that

Keep your thoughts positive
because your thought become
YOUR WORDS.
Keep your words positive
because your words become
YOUR BEHAVIOR
Keep your behavior positive
because your behavior becomes
YOUR HABITS
Keep your habits positive
because your habits become
YOUR VALUES
Keep your values positive
because your values become
YOUR DESTINY
MAHATMA GHANDI

"FAILURE IS NOT AN OPTION". Let students know that you will make math as easy as possible for them but they will have to meet you halfway by continuously thinking.

Step 2: Create an environment where students feel it's okay to struggle. Create activities that are challenging for them. Allow them to struggle and use great questioning to guide their learning. Allow wait time doing students' responses. Different students will need different amounts of wait time. You have to be the judge but the productive struggle is always good.

Step 3: Establish an environment for students to feel safe to be wrong. As we know, some students are reluctant to share answers or thoughts in a math classroom. We have to create a classroom where sharing their thoughts and answers is easily done. As a teacher, I could tell the students that were learning the most because they were the biggest voices in the classes. They would ask questions or either try to answer all the questions. My goal was to get all my students involved. You have to strategically choose students in your classroom. Give the students a voice and even if they are wrong. Guide them to finding the right answer. Don't allow the students to give up but find successes in their answers or responses. Students will be more prone to talk or chime out when they feel safe and encouraged. Create an environment where errors are seen as an opportunity for learning.

Step 4: Change the language of the classroom. When a student says, "I hate math", let them know at the end of this year they will say "I used to hate math". When a child says, "Math is hard, I can't do it", let them know it was hard prior to you being their teacher. It is our job as math teachers to make math relevant, engaging, and understandable to each of our students. You are the voice of the class. Let your voice be heard. By the end of the year, your love for math should shine bright throughout your classroom and students. You should create the path for your students to follow through the world of mathematics.

"Change their mind and the rest will follow."

Relationship Building

"YOU can't teach them, until YOU reach them."

WHEN I WAS GROWING UP, there was not an option, you sit and learn. In most of our homes, education was number one. Even if we did not like school, we had no choice but to be successful. Lets just say, those days have changed. Most students today have to be interested in what you are teaching before they even give you a listening ear. As educators, we wear many hats and if one of your hats is not a relationship builder then the rest of your hats do even matter. In order to truly teach our students, not just going through the motions, we have to reach them. Building relationships between students and teachers are sometimes easily said but it is one of the hardest feats to accomplish in a classroom. Many teachers and students have different ethnicities, cultural backgrounds, and socioeconomic lifestyles. Building relationships can be really challenging dealing with these factors but as educators, we have to go the extra mile. These are some steps I used in my classroom.

Step 1: I never gave work on the first two days of school. I developed an interest survey and an information card to fill out. I filled out the interest survey myself. As a class, everyone would talk about their interest survey. I would start with mine because I wanted to let my students feel comfortable with answering the questions on the survey. I would tell my students about me being scared of the dark growing up just to break the ice. I wanted the students to know I was once a kid with issues myself. Once the students have discussed their surveys, I used the data from the survey to create assignments based on their interests. Also, look at your information cards to learn key information about your students. Some students are being raised by their grandparents, other relatives, single-parent homes, and foster care. These are situations you should be aware of that could impact the learning process. Be intentional about getting to know each of your students. They will care when they know you care.

Step 2: You need to greet your students everyday. Another teacher and I created the 4 H rule for ourselves. The 4 H rule states that you must greet each student with either a hi-five, hug or half-hug, hello or hey, or a handshake. This small token of recognition goes a long way. Students will feel a sense of belonging. Students want to feel loved and welcomed each day. Your greeting may be the one that sets success aside for you and your class for that one student.

Step 3: Every teacher has a set of rules and routines for their classroom. Have your students help develop some of the rules and procedures for your classroom. That gives them ownership in the development of the rules. It gives them a sense of a partnership between themselves and you. Doing this at the beginning of the year will help lay down the foundation for expectations throughout the year. Let it be a two-way street. Allow them to see that you have consequences when you are not fulfilling a part of your role as a teacher. They will see that you are human too.

Step 4: Educators have to create culturally relevant assignments to connect to their students everyday lives. Tyrone Coward states, "teachers must be able to construct pedagogical practices that have relevance and meaning to students' social and cultural realities." When creating word problems in mathematics, develop problems that show a connection between the standards and them. Use real-life examples to illustrate why the math standards are important for them to know. Real world to most may not be real world to a middle schooler that has not left the city they were born in, has not been exposed to opportunities to see much of what the world has to offer through a different lens, or has not had the chance to live a life above middle class and above. Real world means real in their world. Also, use the students' names, apartment complex names, subdivisions or businesses in their neighborhoods in the word problems. Students' eyes light up when they see their names in a problem. They show interest when they are solving something that connects to them. Give students choice when assessing them. Create choice boards and have some of the choices related to their interest. You will get more buy-in from the students.

Building relationships with all your students can be very difficult but as an educator, it is part of your job. It might not be in the job description of the application for teaching you completed but it is one of the most important aspects of teaching. Check the small print, there is a line that refers to unwritten expectations. Consider this is where you will find relationship building with your students. Remember most people will not read a book unless the title or cover page is enticing to them. That goes for students too. If you are not making math interesting for them, they are more prone not to take the time to learn the content.

Mathematical Discourse

"Talk to me, let me know what you are thinking."

PICTURE THIS. YOU JUST TAUGHT the best lesson from beginning to end, at least that's what you thought. For the lesson opener, you told the students about these two cell phone companies. Each company had an attractive option that made their company seem like a better deal. The objective was for the students to determine which company they would select given a specific usage pattern each month. You helped them see how to use the information provided about the companies to write equations for each. You also helped them to understand what each variable in the equations meant, and how to figure out the cost for their usage. What middle schooler doesn't want to save a little money by going with the cheapest company, especially if there is an unlimited usage factor involved. Well, it's now time for the students to work on the "Show What You Know" assignment. This time they are given a situation where they have a specific amount of money for their back to school shopping. Again, another situation where they are vested in how much is spent. As you look around the classroom, you see something very strange and wrong happening. The students are struggling with how to do the work and some can't even get started. They are talking and asking each other questions. Now you, the teacher who just taught this jam up lesson, find yourself at a loss. You start thinking in my head, what don't they understand? "Class, I just showed you all step-by-step what to do." Well, guess what? There lies part of the problem. You showed them, instead of allowing them to discover and find meaning to what they were learning for themselves. Picking topics of interest is good but just isn't enough. Think back to when you were growing up and your parents would say, "you are going to learn your lesson the hard way." Well, it probably wasn't math they were talking about, but when the one thing they tried to get you not to do, you did anyway, and what they tried to warn you about happened. Then is when you finally understood the statement. That was the lesson they referred to when they said "the hard way". That's the kind of learning that students need to help retain the knowledge and apply it in future learning. So where does mathematical discourse come in? I'm glad you asked. Mathematical discourse is the process in which teachers and students communicate their understanding of mathematical concepts. Getting to that understanding is the process or lesson that allows students to gain a deep understanding. This process can be done through conversation or written language but the key is "understanding". So that jam up lesson that was just taught was just a perfect example of one way communication. Now, what does discourse look like and how do you achieve it?

Let's tackle what it looks like first. One of my favorite sayings I used in my math class was "show me your thinking". This could take on many different forms. It could be the thought process for solving an equation, an illustration, chart, figure, or maybe even a graph. There are no limits or restrictions. Any of these will help a teacher tap into the process a student is using to get to the conclusion of a situation or a solution to a problem. Remember, "a picture is worth a thousand words."

You don't always have to start with something tangible. Sometimes a conversation sparks what is needed to organize the thoughts into the process of putting it down on paper. So now, what does mathematical discourse like when there is no paper or pencil being used? This goes to answer the question, how do you achieve it? You, the teacher, will have to be very strategic in your questioning and comments. Instead of answering the student's question with the solution, try answering it with another question that will push their thinking or something that will require them to make a connection. This is how you create back-and-forth dialogue to develop a deeper understanding for learning. Some examples are: "Can you explain your answer?", "Explain how I can isolate the variable.", or "Will multiplying or dividing by a fraction make the answer larger?". When my children were growing up the word "why" was a regular part of their vocabulary. Each time they would ask why, I found myself explaining more and more. Eventually, we finally got to a point of understanding. That same concept applies when the teacher is the one using "why" with intention. The outcome fosters a depth of understanding that will allow students to develop into lifelong learners. Curriculum Associates, LLC has developed a collection of 100 questions and sentence starters that encourage peer and whole class conversations. This resource is designed to support any teacher with being intentional with conversations. There are options for making sense of problems and preserving, reasoning, explaining, critiquing, reflecting, connecting, or just good sentence starters. It is available for free on the world wide web. Just give it a Google.

Effectively using mathematical discourse in the classroom consistently could eventually lead to student learners that will extend their knowledge base beyond your expectations. Here's an example of how using the strategy to introduce a concept can make for a powerful outcome. Please refer to this as a starting point to engage students in effective discourse in your class. Using the strategy regularly promotes a trusting classroom culture, use of effective math vocabulary, and can become a routine expectation for students. Thus, building independent lifelong learners.

SETTING: Students are starting to solve multi-step equations. They have perfected solving one-step equations using all the operations.

Teacher: Today you are going to stretch your thinking on solving equations. We have solved one-step equations using addition, subtraction, multiplication, and division. Those were your baby steps. Now, let's look at how increasing the number of terms in our equation impacts how we arrive at the solution. In other words, no more baby steps. It's time to start speed walking. By the end of the year we will be running a marathon to Algebra. Here is our equation.

$$3x - 22 = 20$$

Teacher: Can someone remind me what our ultimate goal is when we are solving an equation?

Student 1: We want to get the variable by itself on one side of the equation.

Teacher: Which side of the equation and how do I do that?

Student 2: Well, I like to get the variable on the left side but it really doesn't matter.

Student 3: You want to add, subtract, multiply, or divide whatever is needed to get the numbers away from the variable.

Student 4: You mean you want to use inverse operations to isolate the variable.

Teacher: You are both right. That was well said and a very good use of math terms. Help me to figure out which operation I need to do first now that I have more than one in the same equation.

Student 5: I really liked it when there was only one operation I had to decide to use.

Teacher: Well, yes it was easier to decide when there was only one operation. I want you to think about it like juggling a busy schedule on any given day. You can do things in any order you want but it must all get done that day. Now, it would probably make sense to organize them so that you won't be wasting time back and forth and you could make the most of the day. Apply that same concept to solving the equation. Which operation would be the simplest to take the inverse of first?

Student 1: I think adding 22 on both sides of the equation.

Teacher: Why on both sides? I just need it off the side with the variable.

Student 2: You have to add it on both sides to keep the equation balanced.

Teacher: Got it. So, what's next?

Student 3: So the variable is still not by itself. I would divide by 3 on both sides of the equation. That's because 3 is being multiplied by the variable and the inverse is to divide.

Teacher: So $x = 14$. I am finished.

Student 4: Nah, Ms. Smith. You said you always have to substitute your answer back into the equation to check your solution.

Teacher: You are right. Now, what if I decided to divide by 3 before I added 22. Would that change the solution or the steps I have to take to get to the solution?

Student 5: Well, I don't think so.

Teacher: Why don't you come and show us how to do it then.

So, of course, we as teachers know that the detail the student needs to make sure of is to divide all terms by 3 and not just both sides of the equation. This is where students need to be guided through a discussion of why this is the process for multiplication and division in this particular problem. Again, there is so much power in letting students learn through discovery.

The "art" of
TEACHing . . .
is the art of . . .
ASSISTing
DISCOVERY

Building Numeracy

"The bridge between knowing and applying, that's where you will find numeracy."

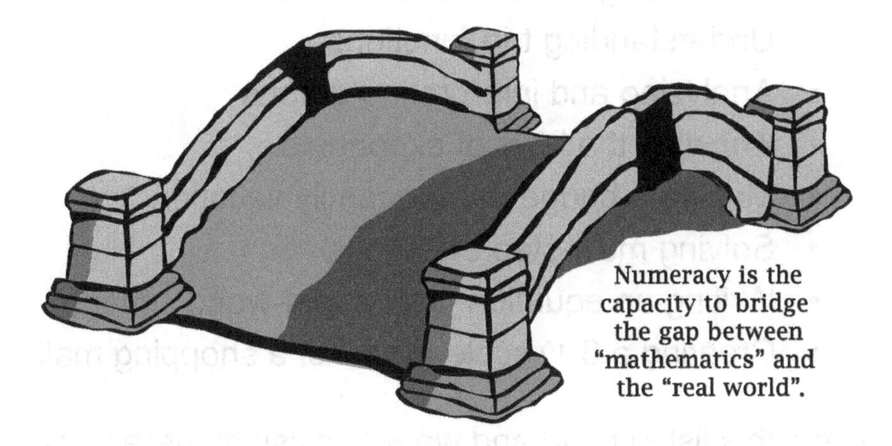

Numeracy is the capacity to bridge the gap between "mathematics" and the "real world".

THERE ARE MANY BUZZWORDS IN the field of education and I am sure you will agree that "numeracy" falls right into that category. Based on your grade band, numeracy can look very different. In the early grades and stages of learning, numeracy and fluency can get a little tangled together. The student focus is on learning the basics and foundational knowledge during these years. As students grow in their learning, numeracy takes on a new understanding and appearance. It moves from knowing the math to applying the math that is learned. For this very reason, teachers have to be intentional when planning and building an environment that supports numeracy for learners. Here is a list of a few things you would see happening in a middle school classroom. You decide if it supports numeracy or is simply learning mathematics.

- Knowing your multiplication facts
- Calculating discounts for a sale
- Understanding trig functions
- Analyzing and interpreting graphs
- Knowing the laws of exponents
- Making a budget for the family vacation
- Solving multi-step equations
- Writing an equation from a real-world situation
- Creating a 3-D scale model of a shopping mall

Keep this list in mind and we will revisit numeracy vs. mathematics at the end of this chapter.

Numeracy is the ability to recognize and understand the role of mathematics in various contexts. It is more than the ability to do basic arithmetic, and it is much more than just knowing the facts. It involves developing confidence and competence with numbers and measures, choosing the mathematics to use, applying mathematical skills, and evaluating their use to solve problems in the world around us. There are a couple of quotes I like to refer to when I help others to understand numeracy vs. mathematics. One is from a family member and is good for sharing with students. He works really hard in helping those he mentors in achieving their fitness goals. Simply said, it is "Process Over Prize", the company slogan for La Rutina Sports & Recreation. This to me says it is more important to dedicate yourself to the steps you have to take to achieve the outcome than it is to focus on the actual outcome. Even closer to home, or should I say to the classroom, our students have to understand the "why" of what they are doing and learning in order to apply it in the real world. The quote that I like to share with adults and teachers might be familiar, "Rome wasn't built in a day." Numeracy doesn't happen overnight or even after one school year. It is a K-12 effort that a student builds upon each year of school. Important work takes time. Learning is very important work.

Students sit in classrooms and put forward their best effort to learn what is taught. How many students really see how to take what they have learned and apply it to their life, their world? You are probably asking, how does one achieve numeracy? What must be done to build a numerate student and when will you know it is achieved? Let's take a look at these details and you can decide how to apply strategies in your class that will meet the needs of your students to achieve this goal.

Students must first understand the importance of numeracy and the role it plays. Teachers must not only "talk the talk" but "walk the walk". It's not enough to just tell a student why they are learning what they are learning, but show them why and how to use what they are learning. This is where "real world" plays such a key role in the learning process. A true understanding of a student's real world is essential. For example, asking students to convert currency from dollars to euros is an example that may be used in a classroom. Students would have to know the conversion factor and apply it. At face value, this is a real-world example, just not their real world. There would be opportunities for some students that have a need but in middle school and at that age it may not be a situation they can connect to right then. When students can see themselves in mathematical real-world situations, they can benefit most. A few examples would be like the cell phone company from the previous chapter, grocery shopping on a budget while comparing unit rates, averaging grades and calculating what is needed to achieve a desired score for the nine weeks. The options are limitless.

Provide students with problem-solving tasks that require them to tap into prior knowledge while discovering the new knowledge desired. Here is an example of a task provided in Big Ideas MATH that requires a student to take what they know and apply it to achieve the desired outcome of writing a math story that could model the data provided.

Essential Question: *How can you use a linear equation in two variables to model and solve a real-life problem? Write a math story using the graph provided.*

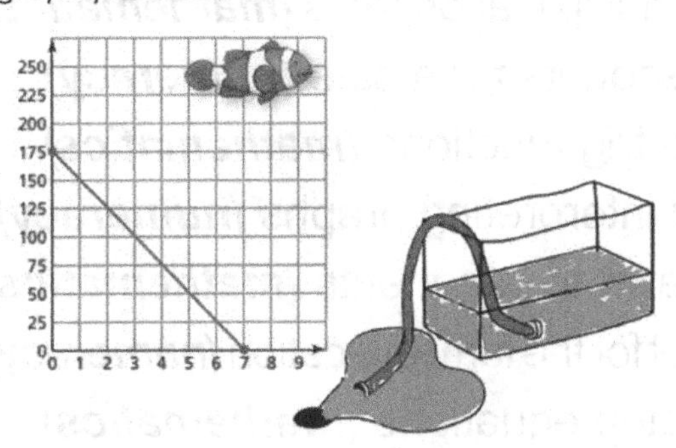

- In your story, interpret the slope of the line, the y-intercept, and the x-intercept.
- Make a table that shows data from the graph.
- Label the axes of the graph with units.

Now, numeracy does not reside only within the four walls of a classroom. A guaranteed way to ensure that students become numerate learners is to build a culture of numeracy in the classroom as well as throughout the entire school. Ways this can be achieved would be to demonstrate that numeracy is everywhere, teachers become facilitators and students are at the center of the learning, inquiry-based learning, developing high quality assessment that are aligned to the intent of the standards, have students explain and justify their solutions, sponsor school-wide competitions designed around numeracy, family math nights, STEM events, and last but certainly not least, take the students on a math walk to explore all that exist around them. Make the learning fun!

So let's revisit the list provided at the beginning of the chapter. You decide, is it numeracy or just plain mathematics?

- Knowing your multiplication facts ***(mathematics)***
- Calculating discounts for a sale ***(numeracy)***
- Understanding trig functions ***(mathematics)***
- Analyzing and interpreting graphs ***(numeracy)***
- Knowing the laws of exponents ***(mathematics)***
- Making a budget for the family vacation ***(numeracy)***
- Solving multi-step equations ***(mathematics)***
- Writing an equation from a real-world situation ***(numeracy)***
- Creating a 3-D scale model of a shopping mall ***(numeracy)***

How did you do?

Differentiation in Math Class

"Change It, Rearrange It, They Will Bring It"

JUST AS MENTIONED IN THE previous chapter, here is another one of those buzz words that seem to take on a new life every time it is mentioned when it relates to education, teaching, and classroom expectations. Here it is, DIFFERENTIATION. I know, you have seen it and heard it so many times. How often have you been asked to make sure you are using strategies for differentiation in your class. Or, here is a better one, you are told that differentiation is something that has to be planned. It doesn't just happen. Well, actually, that second statement is very true.

According to research, differentiated instruction is not one strategy but the approach to what is being taught using a variety of strategies. Carol Ann Tomlison, who is an educator, author, and speaker, communicates it best when she says that differentiated instruction is meeting students' individual needs through content, process, and/or the product for students. So, when the students don't quite get it after three or four examples, that does not mean for you to do three or four more examples the same way. Think about it, they didn't get it the first several times. It also does not mean to do the same thing again slower and louder. What it does mean, those additional examples you have already intentionally included with your planning should be taught using a different strategy. The most valuable advice provided to me very early in my teaching career was to always find out what makes your students "tick". This meant, learn how they learn. Students have unique learning styles. Of course, you can't always teach to address every student's learning style and for every lesson but you can plan with their learning styles in mind so that you know they are being met and the depth of the standard is reaching the student, for the most part.

Let me give you even more food for thought. Have you ever asked yourself, exactly what am I differentiating? Is it the content, the process, or the product? Well, there are strategies for differentiating all three of these. Here are a few strategies that will help with each. Hopefully, it will plant a seed for you to develop even more that will work with your students.

Strategy #1: Learning Stations (Station Rotation) - content or process

This strategy is very popular at the elementary level but is just as effective in middle school, and even high school if done correctly. This strategy can be beneficial in many ways but a culture of respect and expectations must be established in the classroom because students will have to work collaboratively and independently. Each learning station should address the same academic outcome but be achieved using various methods, or multiple representations. A couple of examples of stations for solving systems of equations could be equations to solve, examples of solved equations to do error analysis, word problems to write the equations and solve, and an online virtual activity or game on solving equations.

Strategy #2: Tiered Learning Activities - process

This strategy allows you to meet the needs of students with different ability levels while still covering the intended skill or concept. With this strategy there could be various entry points. A student that is emerging would begin at Level A and work through multiple levels while receiving the additional support needed. A student that is working at an advanced level could enter at Level C where there would be enough to determine they mastered the content and explore extending their knowledge. Planning for this strategy must be intentional.

Strategy #3: Foldables or Graphic Organizers - process

This strategy is ideal for students that are visual learners. Having a tool they can regularly reference helps with their learning process and it allows them to be independent learners as well.

Strategy #4: Manipulatives - process

This strategy is ideal for students that are kinesthetic learners. However, the most important element a teacher must not forget is to make the connection from the concrete to the abstract learning. If this step is missed then the initial learning will be in vain.

Strategy #5: Choice Boards - product

This strategy helps students to own their learning. Students are afforded the opportunity to choose the method in which they show the learned knowledge. They are able to express their creativity while meeting the academic requirement. Examples of choices could be a PowerPoint, collaborative project, class presentation, etc. Again, the end result all connects back to the learning objectives.

Now, one thing I think we can agree upon after reviewing the few strategies shared here is that it will take intentional planning. It will not be an easy task at first. However, it will be a tool in your tool belt that will be sharpened the more you use it. It will get better with time on task.

If a child can't learn
the way we teach,
maybe we should teach
the way they learn.

Ignacio Estrada

From Isolation to Collaboration

"We Are In This Together"

NOW, THAT YOU HAVE GAINED strategies for changing your mindset, relationship building, mathematical discourse and differentiation in the classroom, it is time to break down the walls in the classroom. Even though rooms and buildings have walls, we are not confined to them. In this chapter, you will be given different ways to create a team atmosphere with leadership, parents, and other math educators.

One statement I have heard over the years is "parents are not involved." With all the technology, we have to find ways to entice parents in changing their mindsets about mathematics. One way is to keep them informed. I have used the Remind application on www.remind.com. You can send parents updates about the topics you will be discussing or topics that students need to review. Also, create a webpage for your students or parents to access. You can upload videos of you teaching a concept. You can give the parents the algorithm to solve different problems. With the videos, parents are able to walk the students through different mathematical concepts. If you do not feel comfortable about being videoed, direct parents to youtube videos or different Khan Academy videos. We have to find ways to make parents comfortable with mathematics. Also contact parents periodically throughout the nine weeks. Have an open line of communication with parents. We have to build relationships with parents in order for some of our students to be successful. Many of our students' fears about mathematics are developed from their parents' fears about mathematics.

Also, we have to develop relationships with our colleagues in our schools or within our district. When I first started in education, I felt I was on an island by myself. I was reluctant to ask teachers about strategies because I didn't want to feel like I was incompetent. Over the past years, I realized that I don't have all the answers so I needed to reach out to people. I found that some of my students were struggling with understanding some standards and concepts and I had to take a look in the mirror. "Did I fully understand the standard?" I finally started collaborating with some teachers to get a full meaning of the standards. We developed common formative assessments (CFAs). Common formative assessments are created when two or more teachers collaborate, unpack the standards, and then create assessment questions that will make up the same formative assessments. When teachers create and use CFAs, they are able to analyze data from those assessments and use the data to drive instruction. Below you will find some common formative assessments that were developed by myself and some co-workers.

Collision Course

Rules:

- The teacher places a concept in the middle of the intersection.
- The students have to develop four different strategies, examples or problems that will meet at the common concept in the middle.
- Each of their four examples must incorporate the common concept.
- This activity allows the students to envision how a concept is used in different aspects of a curriculum.

In the example below, students had to create four different types of equations with a solution that equaled the value of 2.

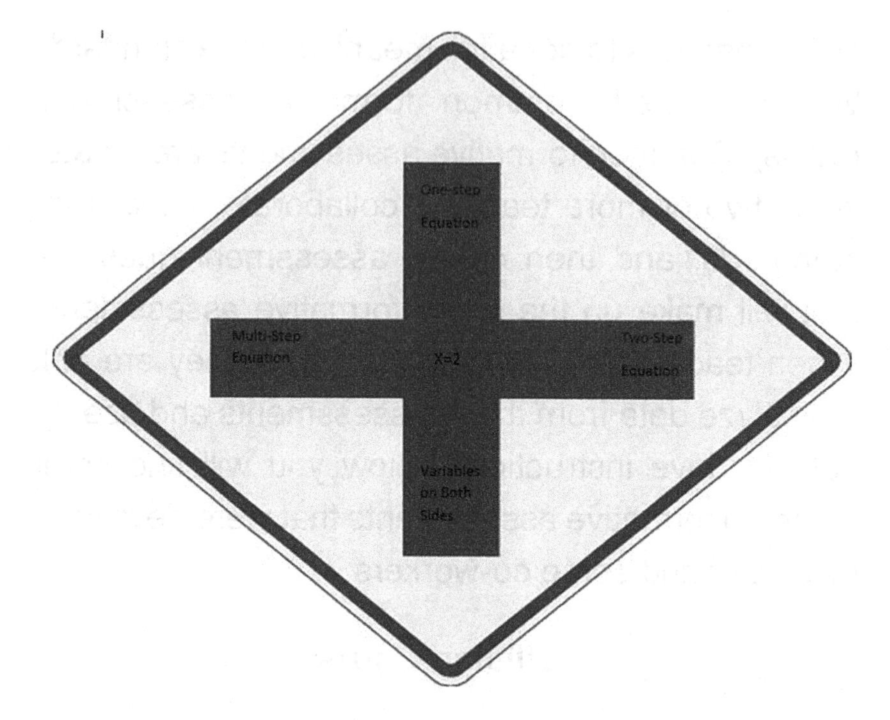

Now that you have created assessments together and analyzed the data, think about strategies that worked or didn't work. If something worked well for another teacher don't be reluctant to equip yourself with some of the tools to put in your toolbox. The overall goal in education is to support all students so if another colleague is doing something differently that works "STEAL IT". We are all in this together. I have stolen or revamped so many of my colleagues' strategies over the years and I am sure they have done the same to me. ***LOL***

Remember we have to move from Isolation to Collaboration. It is essential that we tear down the unseen walls of our classrooms in order to support and engage students. It takes a village to teach a child so we must remove our own barriers to collaborate with the other villagers.

Test Preparation

"Get Ready To Show Your Skills"

IN THE 21ST CENTURY, WE are a test oriented society. Testing starts as early as pre-kindergarten until we start our first job. We have to take academic, agility, ability, social, and any other test someone can develop to show our strengths and challenges. Tests come in a variety of methods. Are we going to know the answer to every question that is asked of us? Are we always going to perform to the highest level asked of us? The answer to those questions is "NO". As a teacher, we are expected to prepare our students for tests. We are required to provide them with the right tools to be successful or do their very best on a given test.

Let's think about this for a moment. As a teacher you spend the entire year teaching students the required standards and the end result is the end of year state assessment. A few weeks before the assessment you do an intense review to ensure that the students have retained what was taught all year. This is the test prep phase. What if we change our mindset about test prep. Instead of it being a few weeks before the assessment, treat everything that is done all year as test prep. One's mindset can set their mind. Awkward statement but it has a powerful meaning. Simply said, your mindset determines (set) how your mind receives and processes the learning. If the thinking is shifted to believing that test prep is the entire year, imagine how students will approach the learning from day 1 to day 180. Everything will be important the first time it is taught and experienced. At a minimum, as a teacher, shifting the mindset for you would mean that planning with the end in mind would make every lesson, assessment, and everything in between be implemented with intentionality.

When we have shifted our mindset, test prep comes to a complete 360. Students will understand that cramming for a final exam or end of course is no longer necessary because they have prepared themselves for the entire year. Teachers will not be faced with the unknowing if their students are prepared for an exam. Teachers will know that the preparation started day one of the year. As a teacher, it is your duty to reflect, rephrase, and reteach on a daily basis so you would not have to take time away from your curriculum to do a month or week long test preparation. We must find different ways to continuously connect standards in mathematics. Students' understanding is necessary for year long test preparation. These are some suggested strategies for year long test preparations.

1. At least once, give the students an exit ticket quick informal assessment on past standards. The ticket to leave can be 2-3 questions based on past knowledge. Use the information from the exit ticket to assess students' previous knowledge and understanding.

2. *Do Nows* or *Bellwork* are questions given at the beginning of class. Teachers use these activities to set the tone of the class and gauge the students' understanding from the previous days' work. Sometimes, I would give the students one question pertaining to a previous standard or lesson. As a class, we would discuss the question and ask how it connects to today's topic. This shows the students the importance of recalling previous knowledge and how it connects to their new learning.

3. From elementary to high school, students love competition and games. This is a great way to review previously learned standards. Students

want to win or be formidable opponents of their classmates. They will study and prepare themselves for the game. Losing is an option but competing to the best of your abilities is not an option.

These are just three ways to do test preparation throughout the year. There are thousands of ways to prepare your students. Google is your best friend. Remember to prepare your students with the end in mind.

Framework For Success

"Foundation is Necessary"

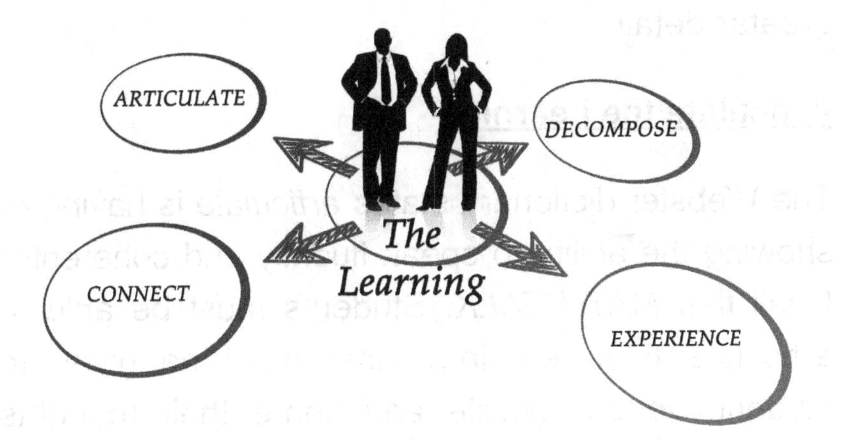

FOR ANY PART OF EDUCATION, you must have a foundation and a framework for any project to be successful. In a mathematics classroom, your curriculum and classroom routines need to be based on some type of framework. The visual above is the framework that I based my curriculum and classroom routines on. There are four essential parts to the framework: articulate the learning, decompose the learning, experience the learning, and connect the learning. Every part of the framework is very important. The parts of the framework are not to be utilized in isolation or in any particular order in the classroom. The model transforms from a traditional classroom to an engaging environment for the students as well as the teachers. Now let us examine each of the parts of the framework in greater detail.

Articulate the Learning

The Webster dictionary states *articulate* is having or showing the ability to speak fluently and coherently. I call this *MATH TALK*. Students must be able to articulate their learning. Time must be given to students to collaborate and share their thoughts about learning. Research has shown, students are more prone to understand if they are able to explain their learning to someone. The learning will be more meaningful to them. Also, it is necessary for students to speak using the correct math terms. They will be more prone to breakdown word problems when they are able to recognize key terms. The bridge between knowing mathematics and understanding numeracy is literacy. We want all our students to be numerate when it comes to mathematics.

Decompose the Learning

Decompose is to breakdown or cause to break apart into simpler constituents. In mathematics, we have to *scaffold learning* for most of our students. The standard is the overall goal or understanding of a topic but many subtopics are embedded in the overarching standing. It would be a perfect world if all our students came with all the prerequisite knowledge of a standard. Since this is not a perfect world, we have to decompose the learning. We have to break it up and apart so that all our students are successful. Our students come to us with different levels of learning, different abilities for learning, and need different ways to learn. As teachers, we must differentiate and scaffold learning so that all our students are successful. They must feel success from the beginning of a lesson to the end of a lesson.

Experience the Learning

Experience means to take part in a situation. The days of lecturing are over. If you go home tired, you are doing the most. Students do not learn by the SIT and GET method of teaching. After a couple of years of teaching, I realized if I went home tired then I did the majority of the work in my class. It is my job to develop activities so my students can be engaged in the learning. Engagement is the key to ensure that students are experiencing the learning. Teachers need to create an engaging environment so students can take part in their learning. Students need to have ownership in their learning. I visited a high school math class and the teachers had different centers set up. The students had to go to each center. The teacher did not teach at all. She told the students the learning goal and gave them instructions on visiting each center. The students went to each center as a group of four. They were able to collaborate and talk about their findings. At the end of the 60 minute class, the teacher was not tired at all. All she did was facilitate and observe the groups at the different centers. At the end of the class, the students spoke as a class about their understanding. We must all students to experience and engage in the learning because it ensures that they have a better understanding of it.

Connect the Learning

Connect has many synonyms that speak volumes such as link, bridge, secure, bind, attach, and join. If a teacher engulfs the meaning of *connect their learning* then they will realize that it is essential for the success of their students. In the world of education, this is better known as real-life application. Students need to see what's the importance of learning a math skill. A teacher needs to find and develop ways to connect math to students' everyday life. A student asked me "Why do we need to know about linear equations?" I told the student let's think about your job at McDonald's. You get paid $8.25 an hour. How many hours do you have to work to be able to get those brand new J's for $200? That is a linear equation ($200 = $8.25h). You don't think of it as a linear equation but it actually is. Even if you go to the grocery store and you purchase apples and oranges, don't you know that is an equation with two variables. If apples cost $0.75 and lemons cost $1.10 an hour, then the equation will be Total Cost = .75A + 1.10L. We have to develop ways to bridge the gap between math standards and real-life application. When students find relevance in learning, they are more prone to take ownership and interest. Always be strategic when connecting math and your students' lives.

The following activity encompasses all facets of the FRAMEWORK.

Upside Down Poster Collaboration Activity:

Material Needed: Posterboard, Two different color markers, a real-life application world problem, an area for both students to work on the poster at the same time.

Directions:

Give each student a copy of the world problem. Each student chooses a different color marker to show their work. Now, tell the students to read the given word problem twice and before they start showing their work on the poster. Draw a line down the middle of the poster because you want their work separate from each other. Each student will do their work in the opposite direction from each other. So the work will be upside down from each other. The students are not able to copy each other's work by working in this manner. After each student has finished, have them turn the board around so they can see their peer's work. Allow the students 3-5 minutes to collaborate and explain their work to each other.

Example: Cat has seven coins in her purse. She only has quarters and dimes. Cat has a total of $1.15. What combination of quarters and dimes does she have to total $1.15?

Tiffany

Quarters + Dimes = 1.15

7 coins = limit

```
  1.15
- .25 Q
  .90
- .10 D
  .80
- .25 Q
  .55
- .10 D
  .45
- .25 Q
  .20
- .10 D
  .10
- .10 D
  0
```

4 dimes

3 quarters

Tim

7 coins = $1.15

Quarters + Dimes Combination

$x + y = 7$

$.25x + .10y = 1.15$

$-.25x + -.25y = -1.75$

$.25x + .10y = 1.15$

$-.15y = -.60$

$y = 4$

$x + 4 = 7$

$-4 \quad -4$

$x = 3$

3 quarters + 4 dimes

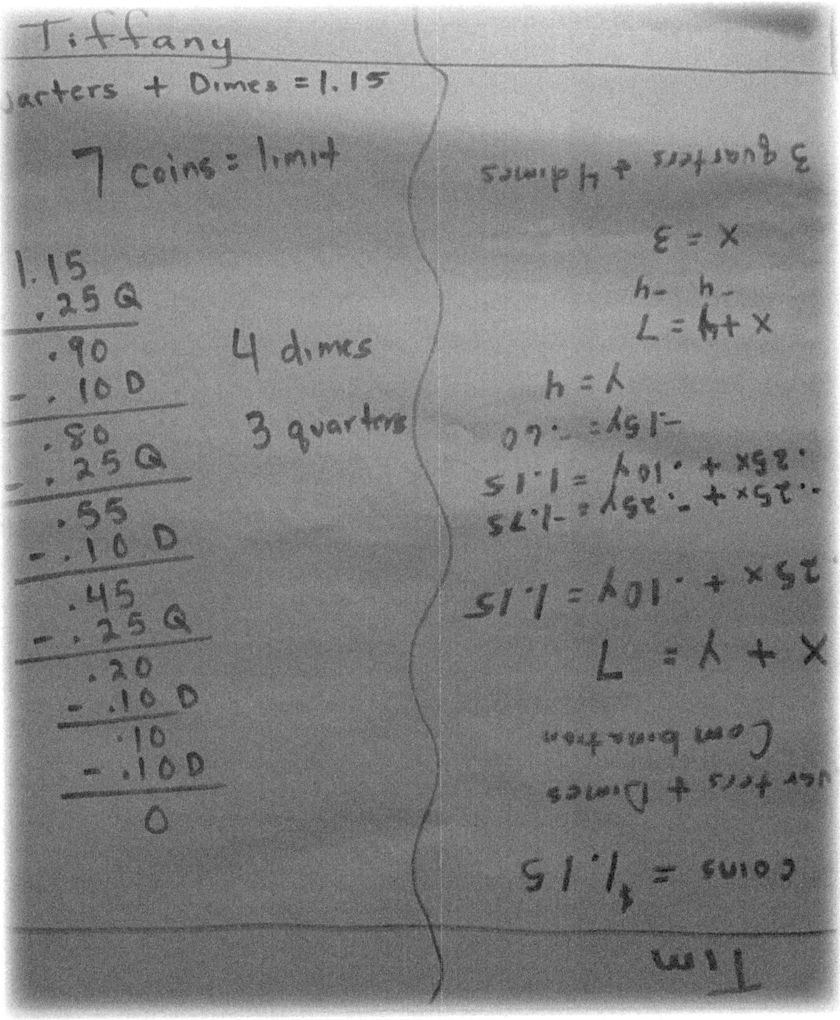

Tiffany

uarters + Dimes = 1.15

7 coins = limit

1.15
-.25 Q
.90
-.10 D
.80
-.25 Q
.55
-.10 D
.45
-.25 Q
.20
-.10 D
.10
-.10 D
0

4 dimes

3 quarters

Tim

coins = $1.15

uarters + Dimes
Combination

x + y = 7

.25x + .10y = 1.15

-.25x + -.25y = -1.75
.25x + .10y = 1.15
-.15y = -.60
y = 4

x + 4 = 7
-4 -4
x = 3

3 quarters + 4 dimes

Assessments

"ACT - Acknowledge, Create, Time"

SOME EDUCATORS SAY THERE IS not a difference between assessing and grading. When a person is assessing a situation, they are in the process of finding ways to improve the situation. When grading, you are in the process of evaluating a situation to give it some type of value. In education, it is necessary to both assess and grade. Our focus in this chapter is to talk about the importance of assessments and the three phases of assessing.

Phase 1

"Acknowledge"

Students should be assessed with some type of assessment activity or tool or a daily basis. Teachers can give informal or formal assessments. When you are informally assessing the students, you are walking around and monitoring their thinking while they are doing independent assignments or you are asking the students questions throughout the lesson. You can use informal assessments as a measuring stick for students' knowledge. Informal assessments are assignments in which you collect or do a quick check. A teacher uses these types of assignments to correct students' misunderstandings. It is our job to collect and analyze data. Now as a teacher, we must acknowledge our students' strengths and weaknesses. We must look at different ways to reteach and differentiate to meet the needs of our students.

Phase 2

"Create"

Once a teacher has acknowledged that differentiating is necessary, it is their job to create lessons, activities, or assignments. These lessons or activities need to be able to enhance their students' understanding. The lessons need to be engaging but reach the intent of the standards. Teachers may need to be the facilitator and not a lecturer. Teachers may need to do center-based activities. These types of activities allow standards to be broken down in multiple ways. They allow teachers to monitor and help facilitate the learning. Students are incharge of their learning and understanding. Students are doing the work instead of listening and writing down multiple examples. Students are able to interact with fellow students. They are able to collaborate as well as communicate with their peers. Also, while students are engaged in the center-based activities, teachers can use this time to work one on one with other students. While a teacher is monitoring students' understanding during the centers, they can pull students and do individualized reinforcement of the standards. There are several activities that a teacher may create in this stage. I recommend centers because they encompass collaboration, differentiation, individualized instruction, and tiered teaching and learning. The most important fact is to ensure that differentiation is considered. Differentiation is not teaching the same lesson at a different pace, louder, or a second time. If a student didn't learn it the first time around, differentiating the delivery allows for another opportunity for learning.

Stage 3

"Time"

How can we find that time to reteach or re assess students? There is a saying, "time waits for no one." This is true. In math, we have so many standards to cover. Sometimes as teachers, we seem so overwhelmed. A lot of times, we think we don't have time to reteach lessons. We have exams and other end of the year assessments and so many deadlines. But as mathematicians, we know that math builds upon itself. You can not move on until students understand what they are covering now. We have to find ways to readjust and allow for reassessing. We have to be very strategic in this manner. I call this the *rob Peter to pay Paul* method. You have to steal time from other standards and show kids how they connect to each other. Use the time to revisit old standards but do not use the time to reteach them. Revisit means to do a quick check. Take about 5-10 minutes to go back over a previous standard. Teachers make the mistake of taking 2 to 3 days to reteach a standard. That much time is not allotted in the pacing guides. Teachers need to do quick revisits on previous standards while teaching other standards throughout the year.

Remember at the end of day assessments and grades are necessary. We have to acknowledge the need of them. Create assignments or activities that are engaging but are to the intent of the standards, and find time to reteach and reassess when students have misconceptions or do not fully understand.

Closure

"The Game Doesn't End Until You Close It Out"

"EVERY TASK, GOAL, RACE
AND YEAR COMES TO
AN END . . .
THEREFORE, MAKE IT
A HABIT TO ALWAYS
FINISH STRONG."

THAT QUOTE SAYS A LOT when it comes to the daily hustle and bustle of your classroom. When it comes to daily lessons, we have so many restraints. We have a set amount of time to teach a concept. Class times can range from 45 to 90 minutes. The students in your class are on various learning levels.

The mathematical standards are so in depth and incorporate so many concepts. How do we teach proficiently when we have so many barriers? How do we close a lesson when we barely have enough time to teach a lesson?

Closure doesn't only happen at the end of a lesson, but can occur throughout the lesson. Closure is the process of checking for understanding within a given lesson. I read an article in Psychology Today, Author Abrigail Brenner (2011) that stated, "**Closure** means finality; a letting go of what once was. Finding **closure** implies a complete acceptance of what has happened and an honoring of the transition away from what's finished to something new." She was talking about moving on in a relationship. Closure in a classroom can also be related to this statement. Once a lesson or part of a lesson is taught, closure will reassure you to move on to the next part. It will be the transition to let you know it is okay to move on to the next. Also, closure can let the instructor know to do some reteaching if needed. There are several techniques to implement closure throughout your lessons. These are just a few.

Quick survey: You can use polleverywhere.com. It is an outline tool used by teachers to check for understanding using polls.

Exit Ticket: This is a quick assessment to have the students do before they leave class.

Check Your Neighbor: This is a peer to peer collaboration strategy. Students explain their learning to each other.

Kahoot It: This is an online educational game. Students can play a quick game to review the lesson.

Back To the Basics: Teachers can do a quiz and gather data to check the level of their students' understanding.

Remember, you cannot complete a task without closure. Closure is essential in your everyday life as well as in the classroom. Never wait until the end, do it throughout your lesson and the ending will be seamless.

Summary

AS AN EDUCATOR YOU WILL cross the paths of many who will offer advice, share strategies, provide demonstrations, and even appear to be an expert at the craft. Remember, we shared very early on, the best lessons you will benefit from are those you will experience. "Living really is learning." That on-the-job training will be invaluable. Learned pedagogy gives you your starting point but the day-to-day with your students and colleagues will help you to really find your way.

We would like to leave you with what we consider some of the most important take-aways.

- Be your own artist. There is nothing wrong with taking a strategy and delivering it with your flare and flavor. Don't be discouraged if the results are different. That is to be expected. Make it your own!

- Start out with small steps. Taking on too much at one time may not yield the desired results. Implement two to three strategies that will make the biggest impact on learning with your students and perfect them before adding more into your classroom routine. This will allow you to see what works, what needs tweaking, and what needs to be placed in the file cabinet for another time.
- Remember to communicate and collaborate with your colleagues. Every educator needs a strong support system. There are always people that are willing to lend a helping hand. Do not try to fight the battle by yourself. Remember just about all superheroes had to count on someone else from time to time. And please don't forget about the support a parent can provide. Consider them a secret weapon.
- Don't get burnt out. Taking time out for self is the key to longevity when working in the field of education. When you are struggling or worn out mentally or physically just take a breath and step away for just a moment.
- Mindset leads the way. A growth mindset can open doors for new learning that you can only imagine if you let it. A fixed mindset stifles your creativity but a growth mindset is the key to limitless possibilities.

In essence, it is our hope and desire that this resource provided you just enough to ignite a spark that started that flame in your thoughts and actions to be the very best educator you desire for your students.